Kindergarten Skills

MW00610647

Table of Contents

ISBN 978-1-4838-0511-5

02-243151151

Ideas for How to Use This Book

This book was developed to help children master basic skills that form the foundation for children's learning. The stronger their foundation is in the basics, the easier children will be able to progress through more challenging tasks.

All children learn at their own rate; therefore, introduce concepts to children when developmentally appropriate. In addition, provide opportunities for hands-on learning to reinforce the skills covered within the activity pages.

The back of this book includes removable flash cards to use for basic skill practice and enrichment activities. Pull the pages out and cut the flash cards apart.

The following are some ideas for using these flash cards. It is recommended that you begin each activity using only a few cards. Gradually introduce cards as children become more proficient.

- Mix the 26 uppercase letter flash cards and put them in a stack. Have children put them in alphabetical order. When finished, say the letters of the alphabet together.

- Spread the 26 uppercase letter flash cards out on a flat surface. Hold the 26 lowercase cards and show one card at a time. Have children match the lowercase letter to the correct uppercase letter. Discuss any similarities and differences each pair of letters might have.

- Use the 26 lowercase letter flash cards to show children one letter at a time. Ask them to name each letter and to provide a word that begins with the same letter. Write each word for children to see.

- Using the numeral flash cards, have children put the flash cards in numeric order from one to fifty. When finished, invite them to use the flash cards to count to fifty.

 CD-104638 • © Carson-Dellosa

Ideas for How to Use This Book

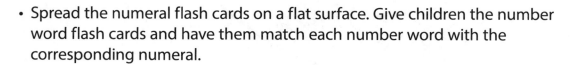

- Have children place the even numeral flash cards into one group and the odd numeral flash cards into a second group. Encourage them to use the flash cards to count by odds and then by evens.

- On a flat surface, spread the math signs for *greater than*, *less than*, and *equals*. Then, show children two numeral cards at a time. Have children choose the correct math sign.

- Gather the numeral flash cards, the flash cards with math signs, and objects for counting, such as buttons, paper clips, beans, or pennies. Show children two numeral flash cards at a time. Have them show each number with objects. Work with children to use the math sign cards to set up simple equations.

- Spread the numeral flash cards on a flat surface. Give children the number word flash cards and have them match each number word with the corresponding numeral.

- Show children one colorful shape flash card at a time. Have them identify each shape and its color. Encourage children to name real-life objects that are the same shape.

- Spread the colorful shape flash cards on a flat surface. Give children the shape word flash cards and encourage them to match each shape word to the correct shape. Discuss objects around the room that are the same color as each shape.

- Turn the shape word flash cards and the colorful shape flash cards over on a flat surface. Encourage children to play a game of concentration by matching each shape with the correct shape word.

Name _____

Following a Maze

Follow the maze to help the butterfly find the flower.

CD-104638 • © Carson-Dellosa

Name _____

Following a Maze

Follow the maze to help the dog find the bone.

Name _____

Following a Maze

Follow the maze to help the horse find the barn.

Name _____

Following a Maze

Follow the maze to help the cat find the ball of yarn.

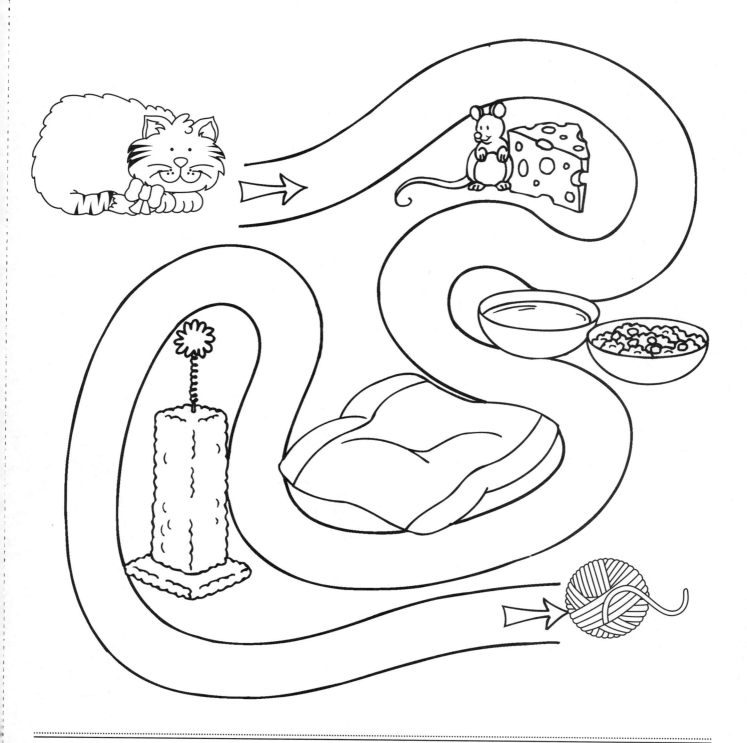

Name _____

Tracing Straight Lines

Trace the dotted lines from **left** to **right** to help each mouse get to its cheese.

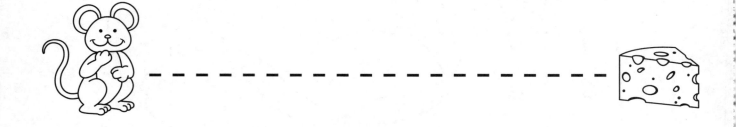

Name _____

Tracing Straight Lines

Trace the dotted lines from **top** to **bottom** to help each bee find its hive.

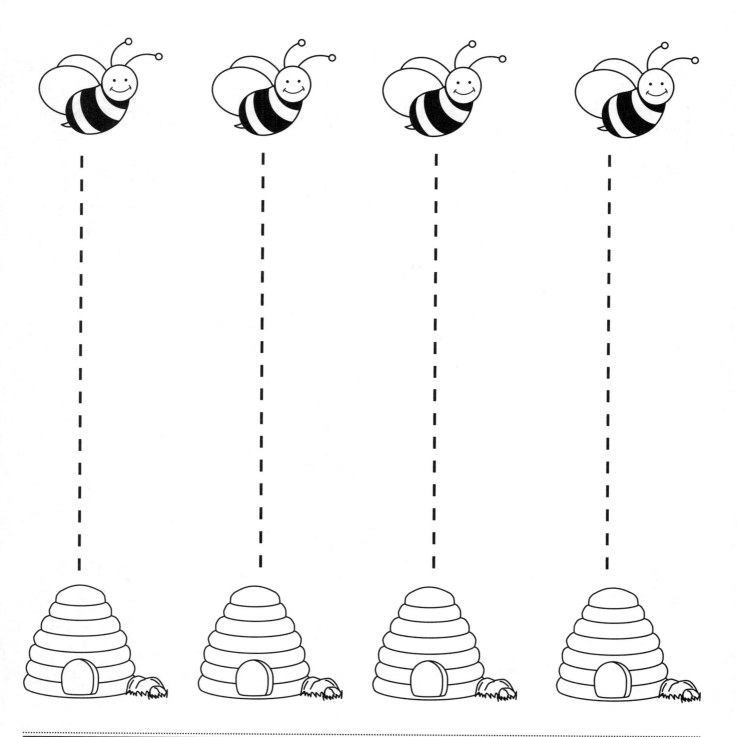

Tracing Straight Lines

Trace the dotted lines from **top** to **bottom** to help each school bus get to school.

Name _____

Tracing Straight Lines

Trace the dotted line to help the rocket find its way to the moon.

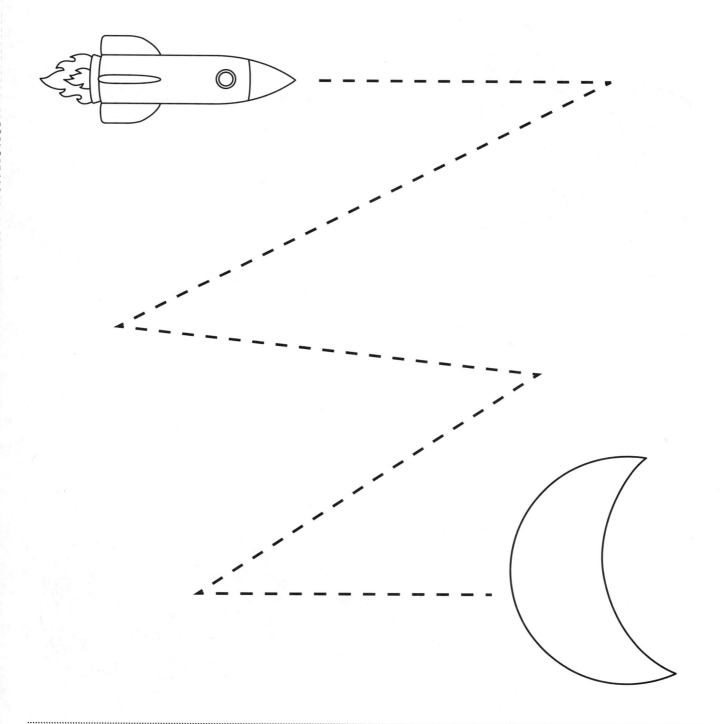

Tracing Curved Lines

Trace the dotted line to help the bird find its nest.

Tracing Curved Lines

Trace the dotted lines to help each rabbit hop to its carrot.

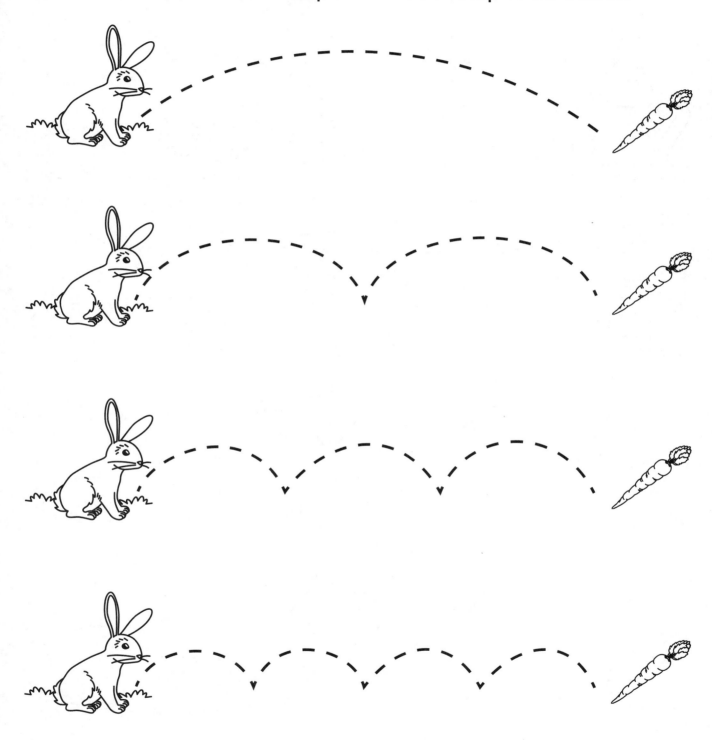

Tracing Curved Lines

Trace the dotted lines to help each frog hop to its lily pad.

CD-104638 • © Carson-Dellosa

Name _____

Tracing Curved Lines

Trace the dotted lines to help each boy get to his island.

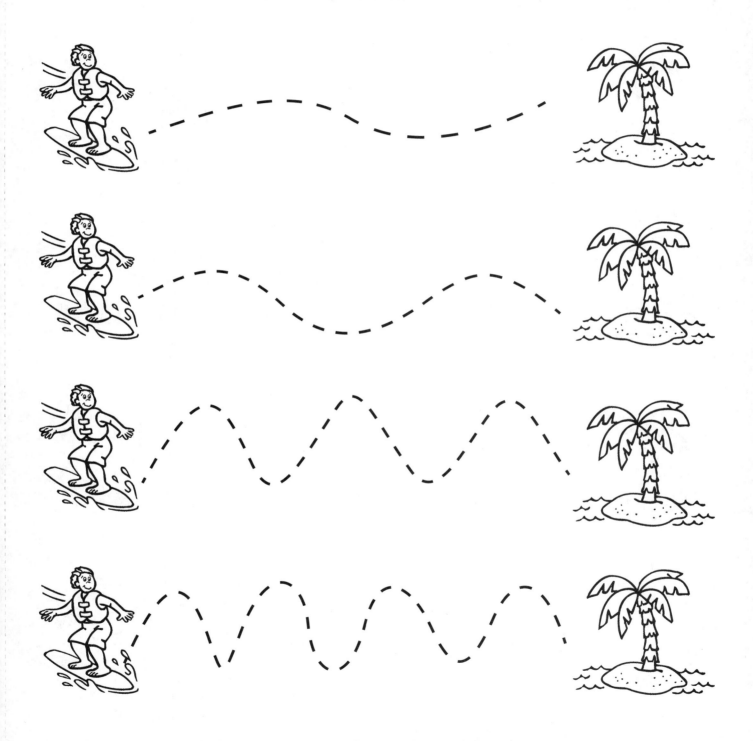

Recognizing Same and Different

Draw a line to match the pictures that are the **same**.
Color the matching pictures the same color.

Name _____

Recognizing Same and Different

Draw a line to match the pictures that are the **same**.
Color the matching pictures the same color.

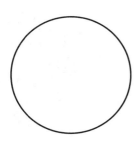

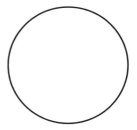

Recognizing Same and Different

Circle and color the pictures in each row that are the **same** as the first picture.

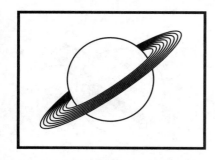  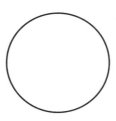 

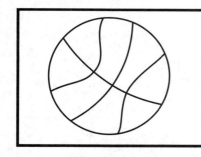   

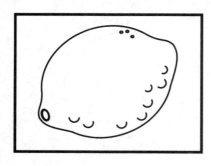   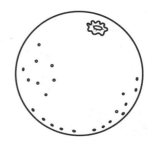

Recognizing Same and Different

Circle and color the pictures in each row that are the **same** as the first picture.

  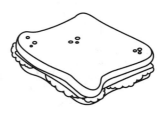 

 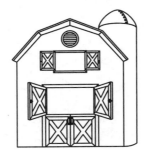  

Name _____

Recognizing Same and Different

Cross out the picture that is **different** in each row. Color the pictures that are the **same**.

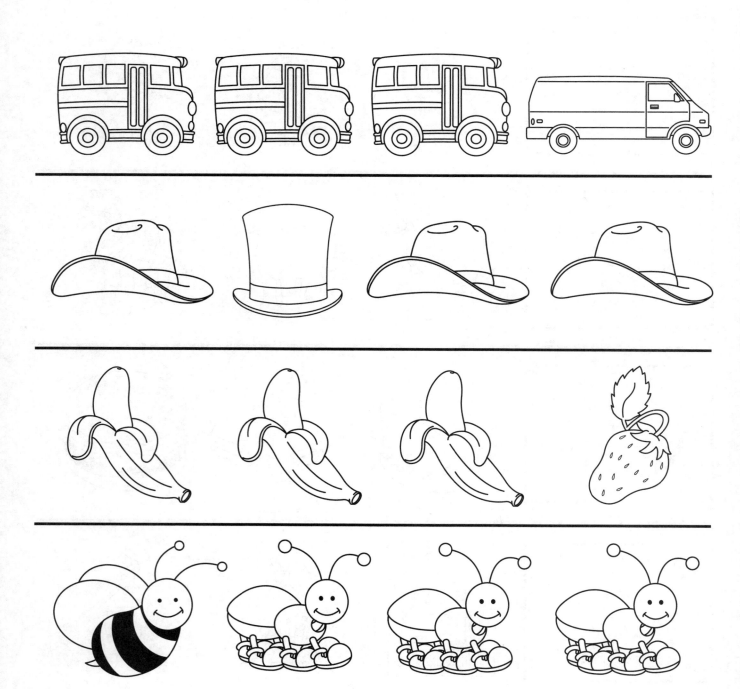

CD-104638 • © Carson-Dellosa

Name _____

Recognizing Same and Different

Cross out the picture that is **different** in each row. Color the pictures that are the **same**.

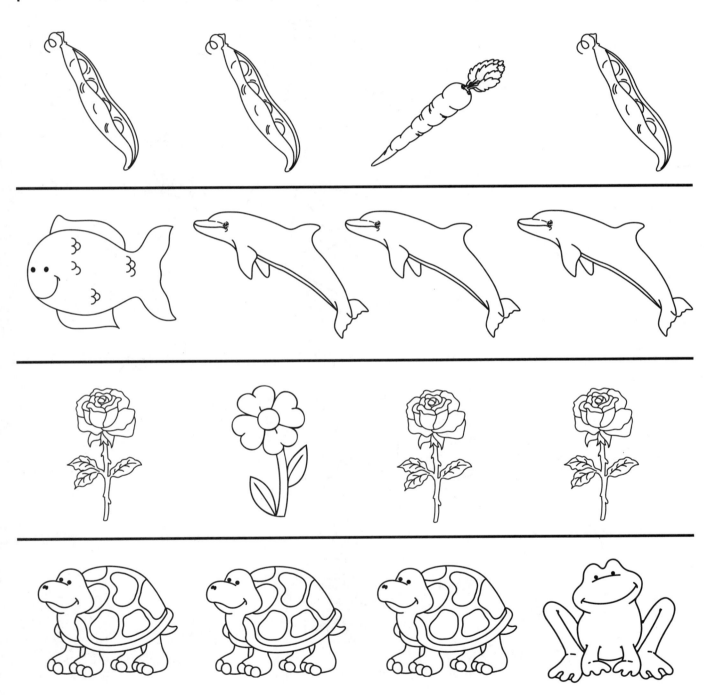

Name _____

Classifying Clothes

Circle and color the pictures of clothes.

Name _____

Classifying Animals

Circle and color the pictures of animals.

Name _____

Classifying Food

Circle and color the pictures of food.

CD-104638 • © Carson-Dellosa

Recognizing Big and Small

Color the **big** kites red. Color the **small** kites blue.

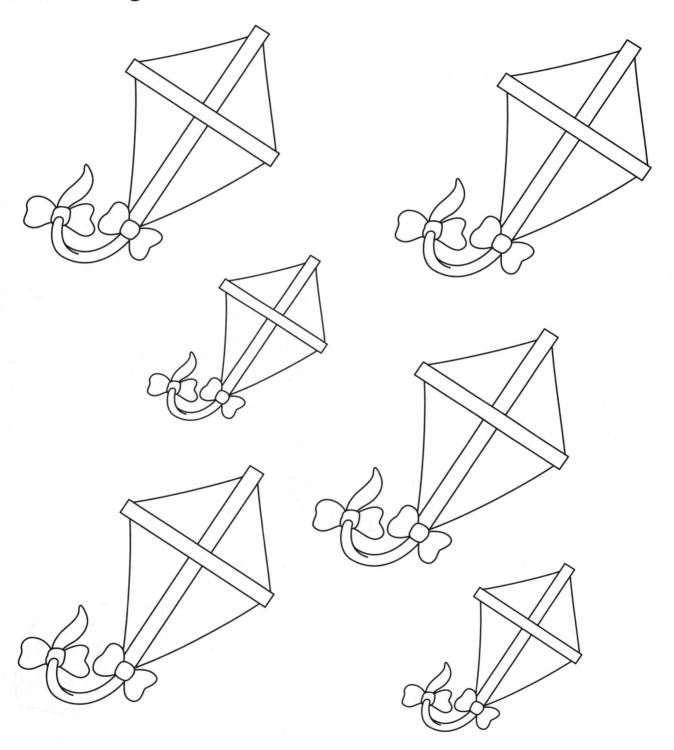

Name _____

Recognizing Big and Small

Color the **big** gifts orange. Color the **small** gifts yellow.

Recognizing Big and Small

Color the **big** fish green. Color the **small** fish yellow.

Name _____

Recognizing Long and Short

Color the **long** carrots orange. Color the **short** carrots yellow.

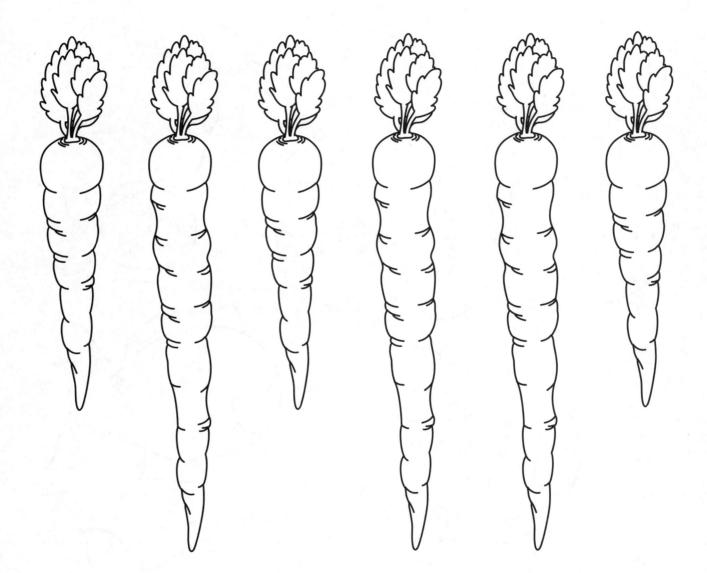

Recognizing Long and Short

Color the **long** pencils green. Color the **short** pencils orange.

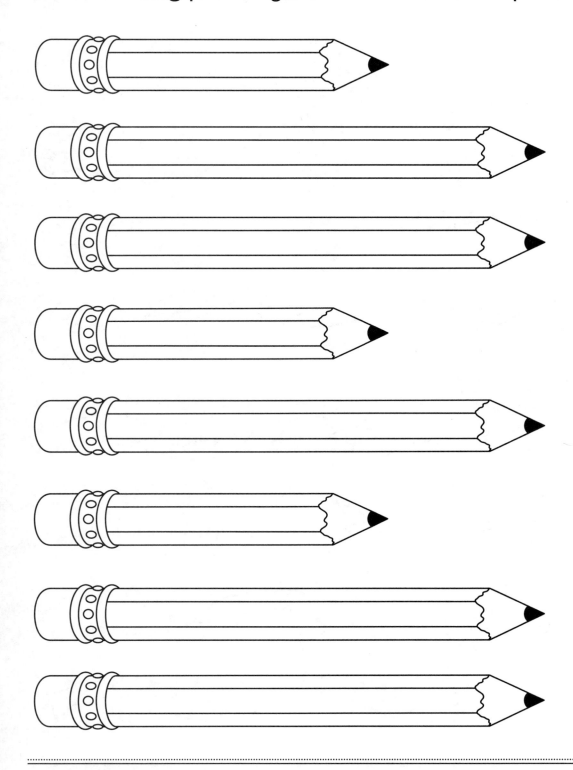

Recognizing Long and Short

Color the **long** pieces of rope red. Color the **short** pieces of rope purple.

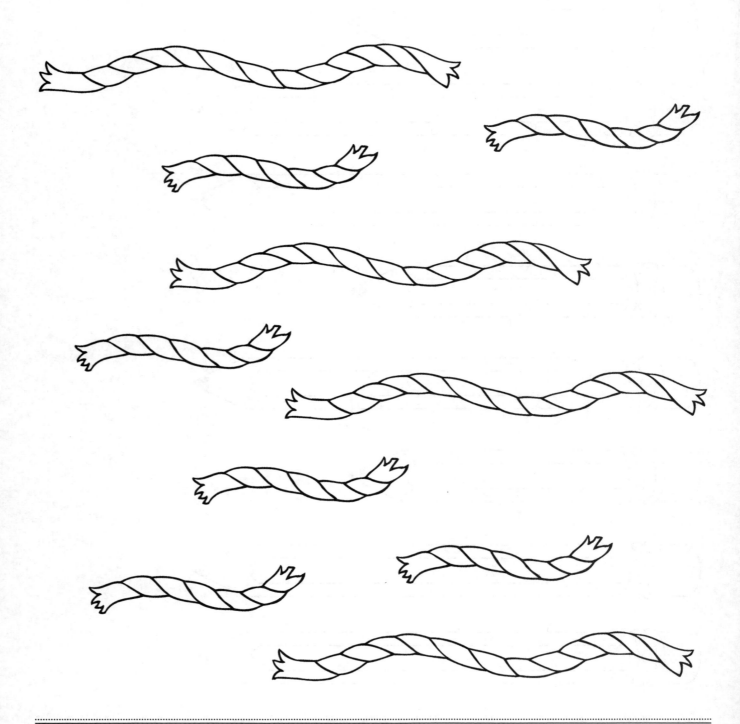

Recognizing Largest

Circle and color the **largest** object in each row.

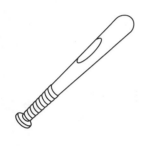   

 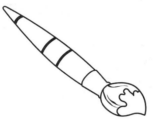  

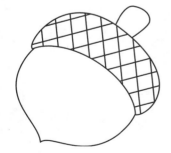   

Name _____

Recognizing Largest

Circle and color the **largest** object in each row.

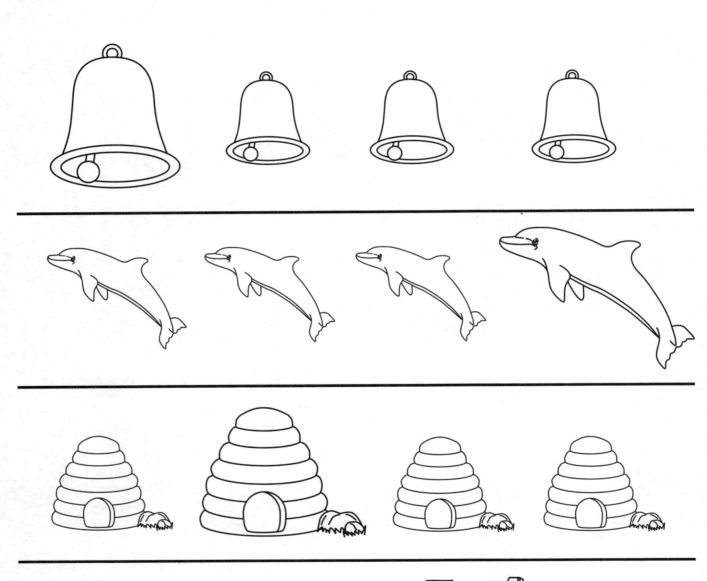

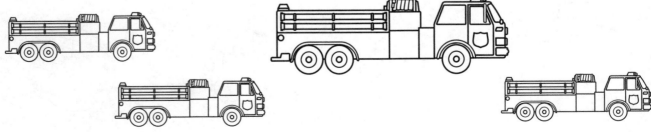

Recognizing Smallest

Circle and color the **smallest** object in each row.

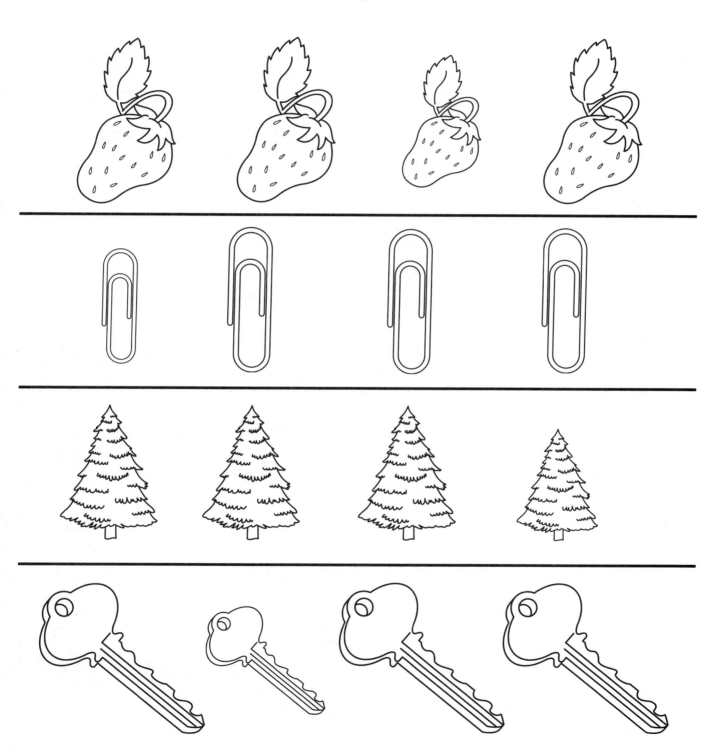

Name _____

Recognizing Smallest

Circle and color the **smallest** object in each row.

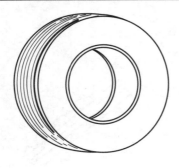  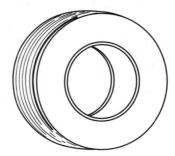 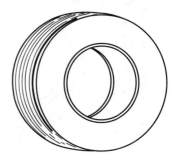

CD-104638 • © Carson-Dellosa

Name _____

Cut and Paste a Car

Cut out and paste the objects onto the matching spaces to complete the picture. Color the picture.

Cut and Paste a Butterfly

Cut out and paste the objects onto the matching spaces to complete the picture. Color the picture.

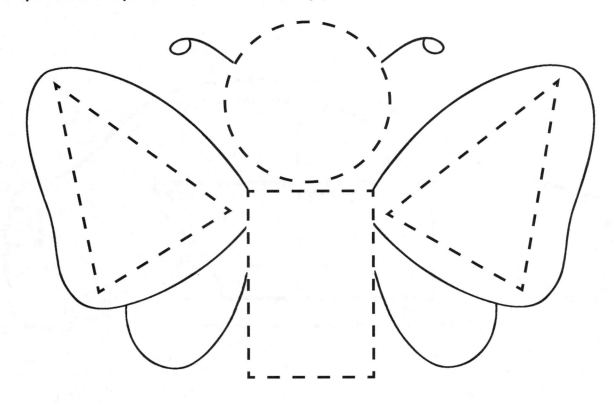

Name _____

Cut and Paste Flowers

Cut out and paste the objects onto the matching spaces to complete the picture. Color the picture.

  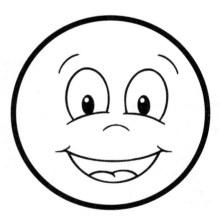

Cut and Paste Windows

Cut out and paste the objects onto the matching spaces to complete the picture. Color the picture.

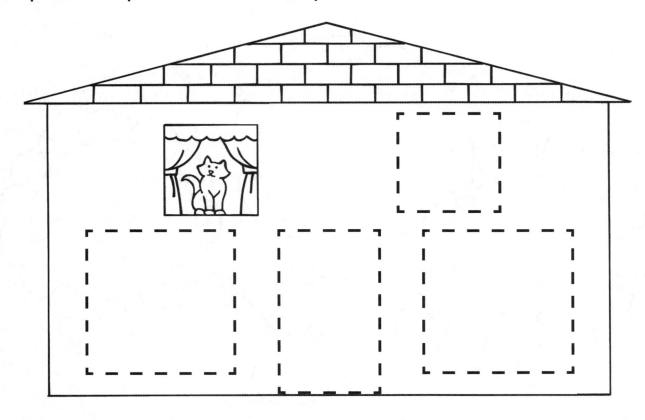

  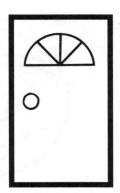 

Name _____

Cut and Paste Pyramids

Cut out and paste the objects onto the matching spaces to complete the picture. Color the picture.

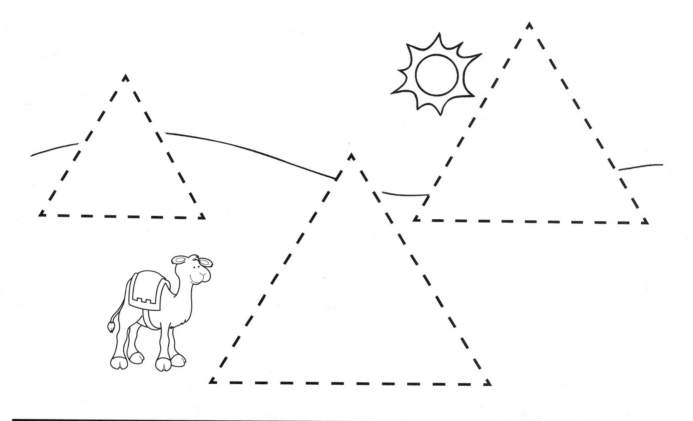

Name _____

Writing the Letters Aa and Bb

Practice writing the letters.

 ant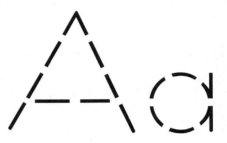

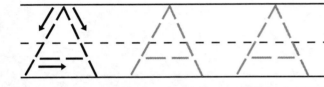

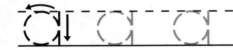

 bee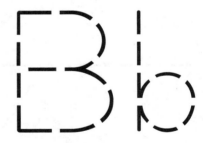

 CD-104638 • © Carson-Dellosa

Writing the Letters Cc and Dd

Practice writing the letters.

cow

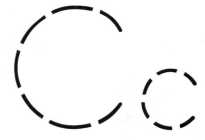

dog

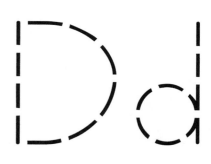

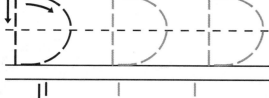

Writing the Letters Ee and Ff

Practice writing the letters.

elephant

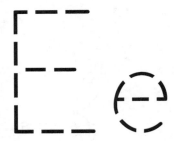

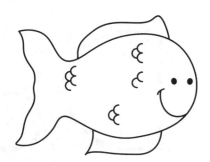

fish

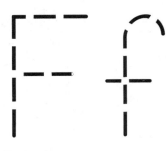

Name _____

Writing the Letters Gg and Hh

Practice writing the letters.

 goat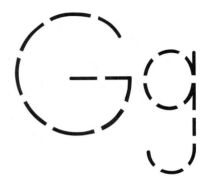

 hat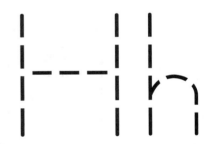

Name _____

Writing the Letters Ii and Jj

Practice writing the letters.

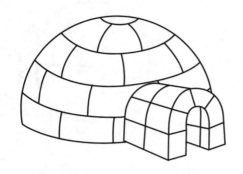 igloo

 jam

CD-104638 • © Carson-Dellosa

Writing the Letters Kk and Ll

Practice writing the letters.

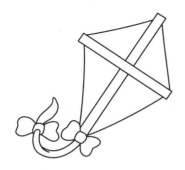

kite

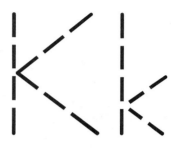

lion

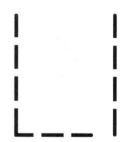

Writing the Letters Mm and Nn

Practice writing the letters.

mouse

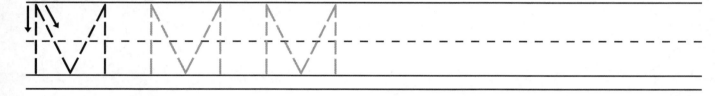

nest

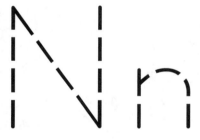

CD-104638 • © Carson-Dellosa

Writing the Letters Oo and Pp

Practice writing the letters.

 octopus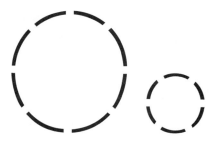

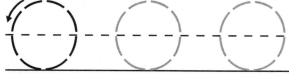

 pig

Name _____

Writing the Letters Qq and Rr

Practice writing the letters.

 queen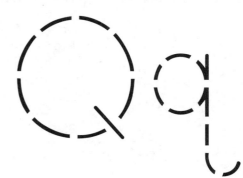

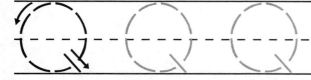

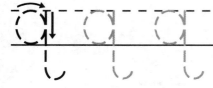

 rabbit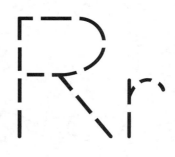

Name _____

Writing the Letters Ss and Tt

Practice writing the letters.

sun

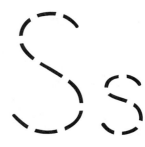

S S S

S S S

turtle

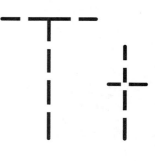

T T T

t t t

Writing the Letters Uu and Vv

Practice writing the letters.

umbrella

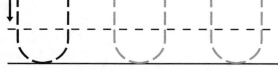

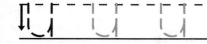

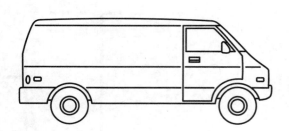

van

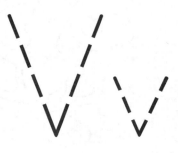

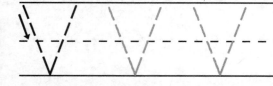

Writing the Letters Ww and Xx

Practice writing the letters.

watch

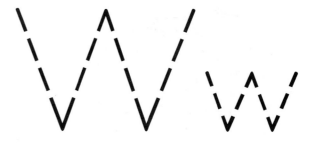

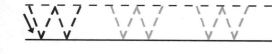

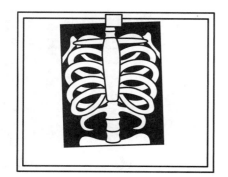

X-ray

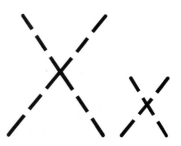

Writing the Letters Yy and Zz

Practice writing the letters.

yo-yo

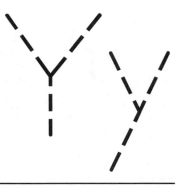

zebra

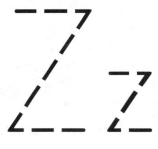

Writing Uppercase Letters

Trace the dotted lines to complete each **uppercase** letter.

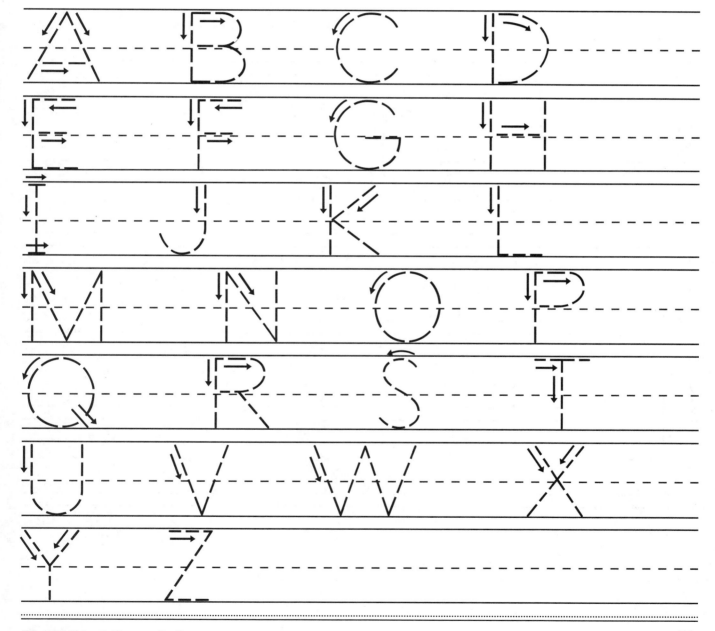

Writing Lowercase Letters

Trace the dotted lines to complete each **lowercase** letter.

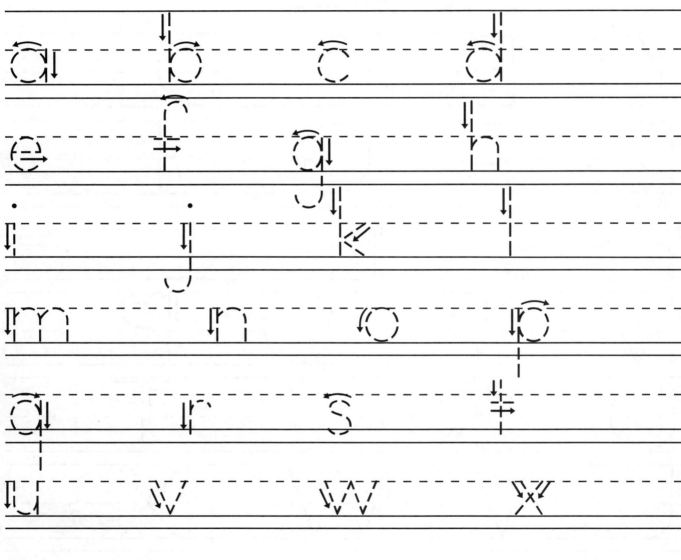

Name _____

Writing and Ordering Uppercase Letters

Fill in the blanks with the missing **uppercase** letters.

A _ _ C D _ _ F

G H I _ _ _ _ _ L

_ N _ _ _ _ Q R

S _ _ U V _ _ X

_ _ Z

Writing and Ordering Lowercase Letters

Fill in the blanks with the missing **lowercase** letters.

a __ __ __ d e __

g __ i __ k l

m __ o p __ __ __

s t __ __ v __ __ x

y __ __

Dot-to-Dot Uppercase Letters

Connect the dots by following the **uppercase** letters of the alphabet. Begin with **A**.

Dot-to-Dot Lowercase Letters

Connect the dots by following the **lowercase** letters of the alphabet. Begin with **a**.

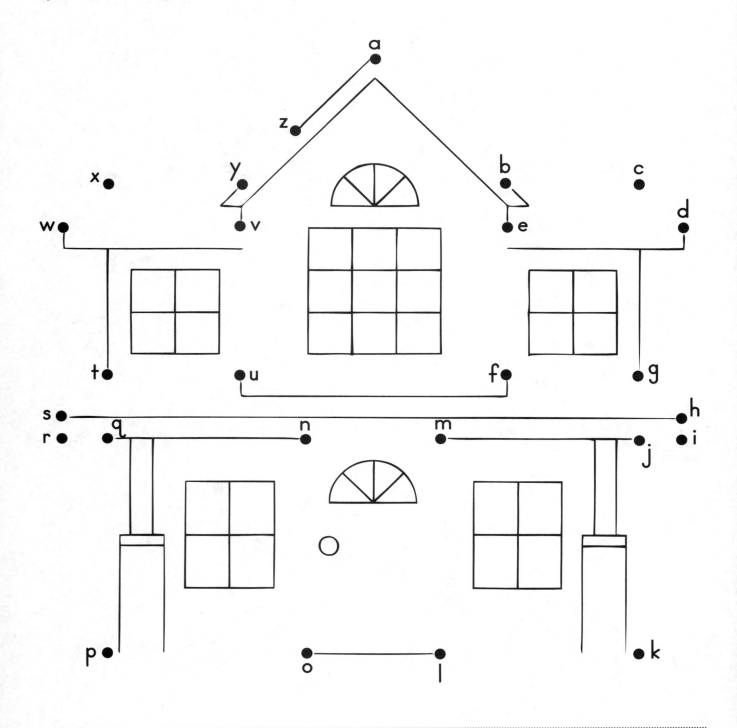

Name _____

Matching Uppercase and Lowercase Letters

Draw a line to match each **uppercase** letter with the correct **lowercase** letter.

Name _____

Matching Uppercase and Lowercase Letters

Draw a line to match each **uppercase** letter with the correct **lowercase** letter.

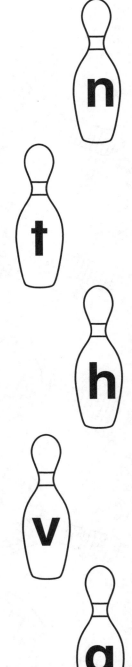

Matching Uppercase and Lowercase Letters

Draw a line to match each **uppercase** letter with the correct **lowercase** letter.

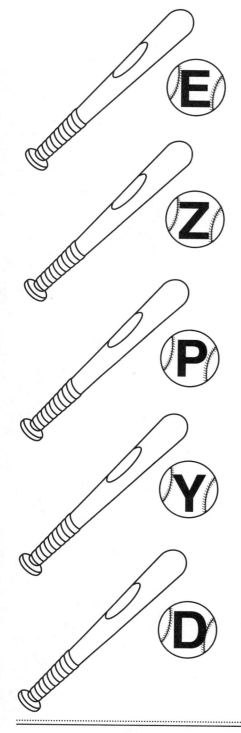

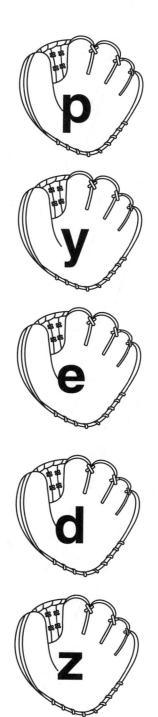

Name _____

Matching Uppercase and Lowercase Letters

Draw a line to match each **uppercase** letter with the correct
lowercase letter.

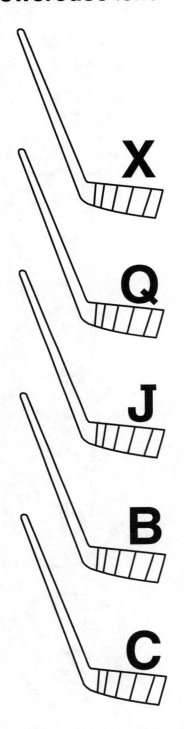

X

Q

J

B

C

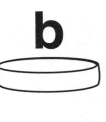

b

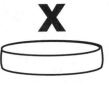

x

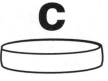

c

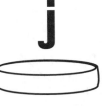

j

q

Matching Uppercase and Lowercase Letters

Draw a line to match each **uppercase** letter with the correct **lowercase** letter.

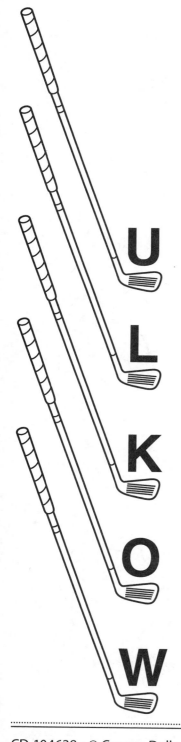

U

L

K

O

W

I

w

u

k

o

Name _____

Learning Red

Use a **red** crayon to color the things that are usually **red**.

CD-104638 • © Carson-Dellosa

Name _____

Learning Orange

Use an **orange** crayon to color the things that are usually **orange**.

Name _____

Learning Yellow

Use a **yellow** crayon to color the things that are usually **yellow**.

CD-104638 • © Carson-Dellosa

Learning Green

Use a **green** crayon to color the things that are usually **green**.

Name _____

Learning Colors

Color the pictures the correct color.

blue bird

brown acorn

red apple

green turtle

yellow star

CD-104638 • © Carson-Dellosa

Learning Colors

Color the pictures the correct color.

pink flamingo

black bat

orange pumpkin

gray elephant

purple grapes

Name _____

Hidden Picture: Green and Brown

Color each space the correct color. Finish coloring the picture.

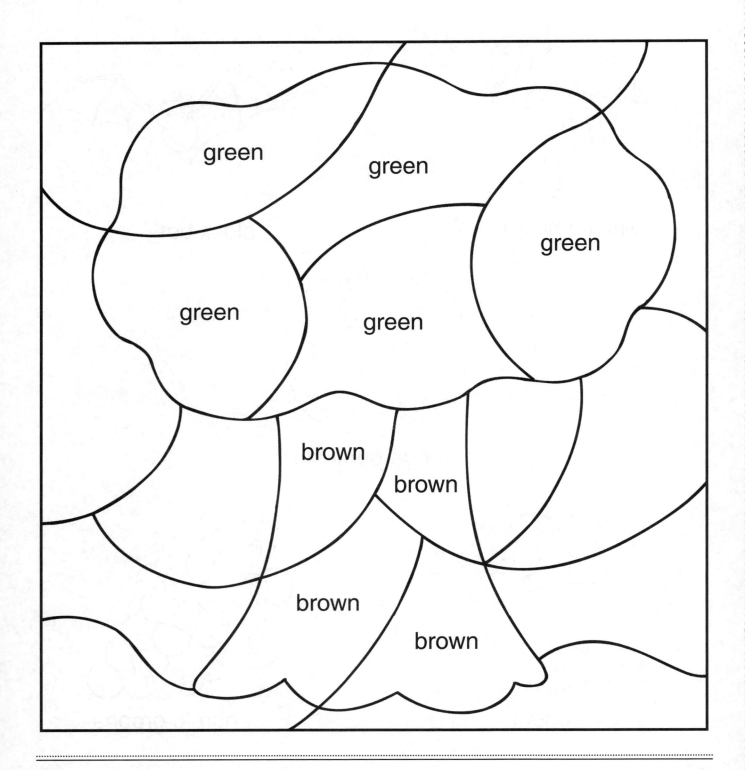

Name _____

Hidden Picture: Orange and Black

Color each space the correct color. Finish coloring the picture.

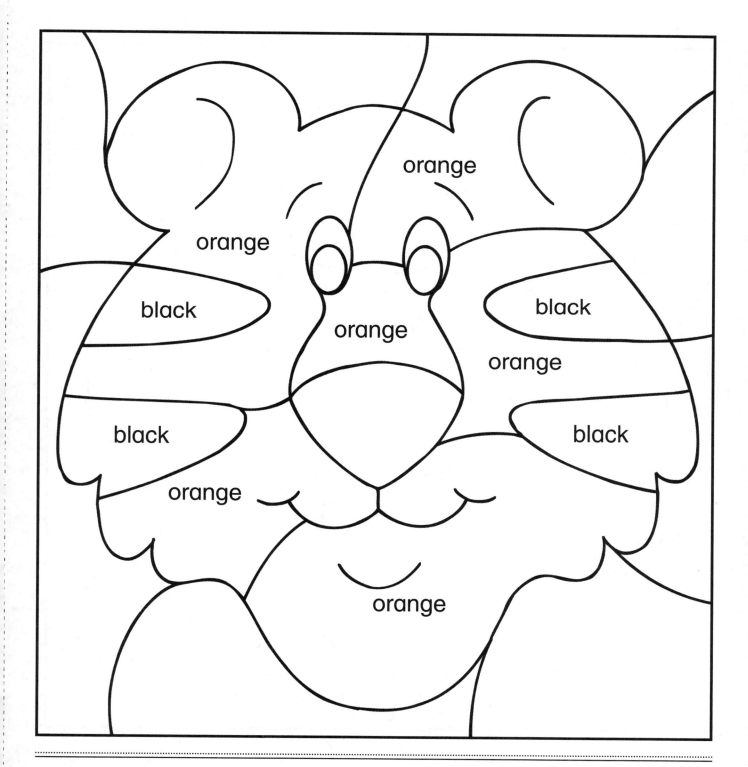

Hidden Picture: Red and Pink

Color each space the correct color. Finish coloring the picture.

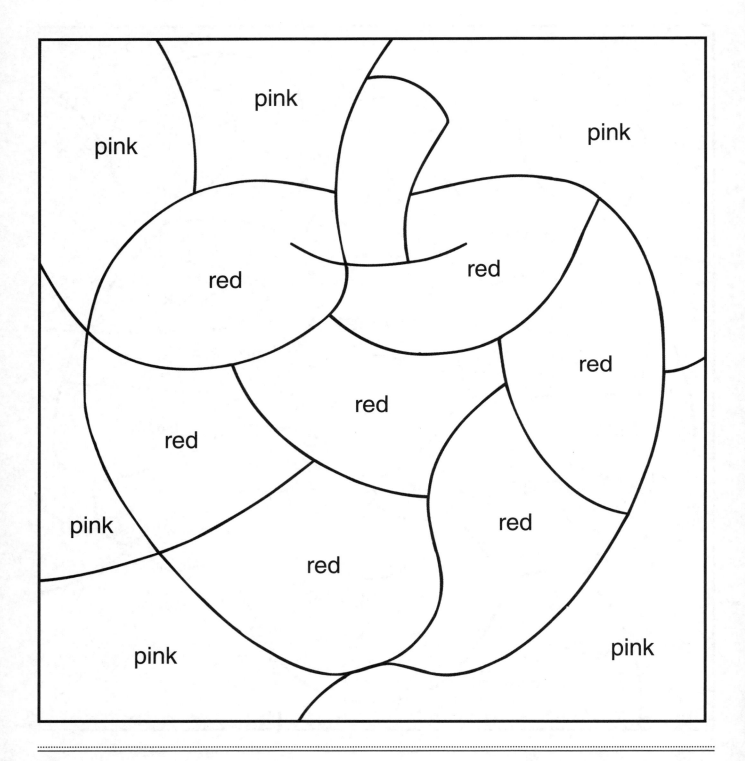

Hidden Picture: Yellow and Blue

Color each space the correct color. Finish coloring the picture.

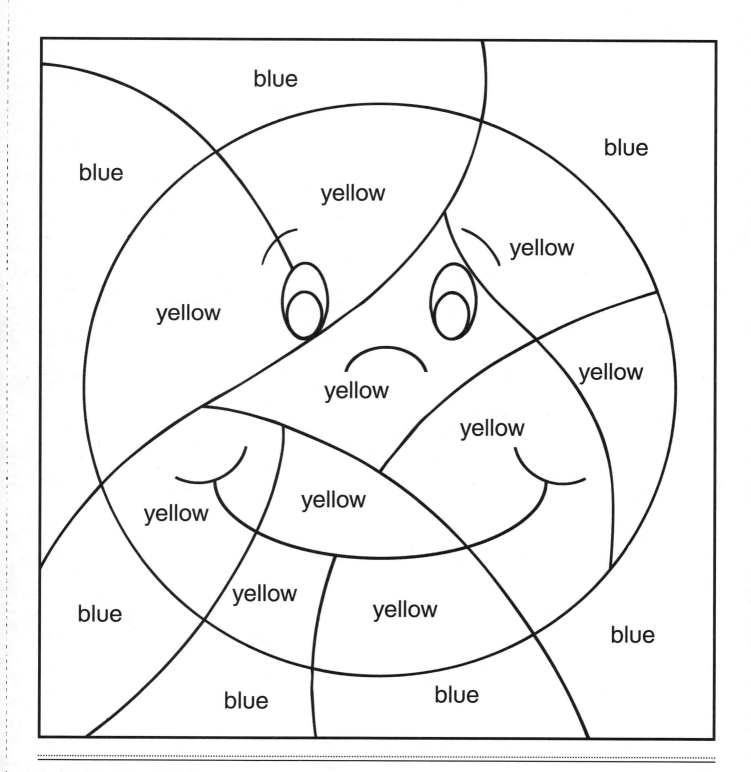

Name _____

Writing and Counting to One

Count and color the moon. Practice writing the numeral and number word.

one moon

Circle and color one star.

CD-104638 • © Carson-Dellosa

Name _____

Writing and Counting to Two

Count and color the queens. Practice writing the numeral and number word.

two queens

2 2 2 _____

two two _____

Circle and color two crowns.

Name _____

Writing and Counting to Three

Count and color the dogs. Practice writing the numeral and number word.

 three dogs

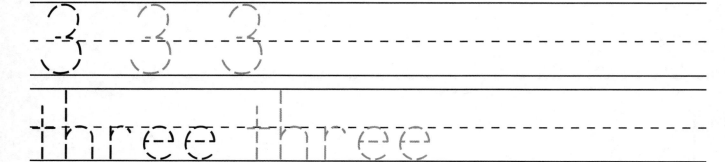

Circle and color three bones.

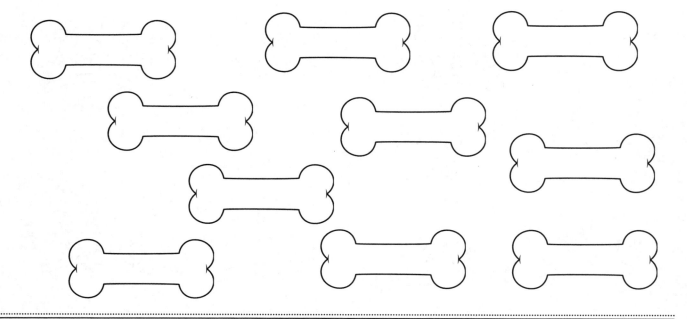

Name _____

Writing and Counting to Four

Count and color the cows. Practice writing the numeral and number word.

four cows

4   4   4 _____

four four _____

Circle and color four barns.

Name _____

Writing and Counting to Five

Count and color the boats. Practice writing the numeral and number word.

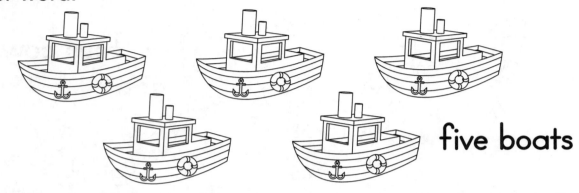

five boats

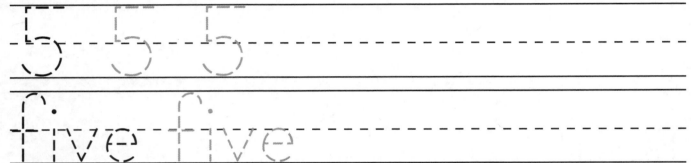

Circle and color five fish.

CD-104638 • © Carson-Dellosa

Writing and Counting to Six

Count and color the igloos. Practice writing the numeral and number word.

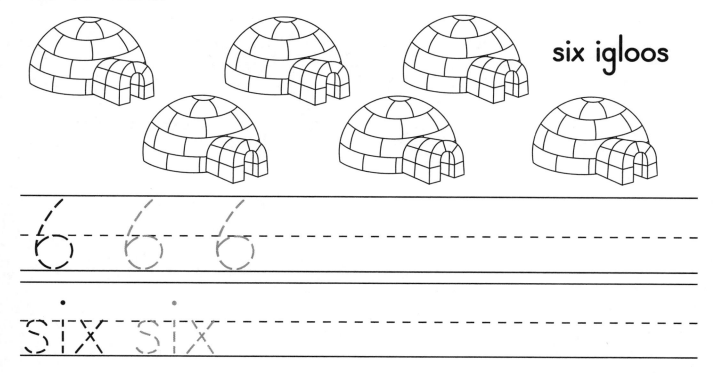

six igloos

Circle and color six coats.

Name _____

Writing and Counting to Seven

Count and color the nests. Practice writing the numeral and number word.

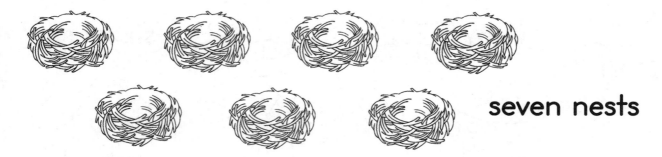

seven nests

7 7 7

seven seven

Circle and color seven birds.

CD-104638 • © Carson-Dellosa

Name _____

Writing and Counting to Eight

Count and color the webs. Practice writing the numeral and number word.

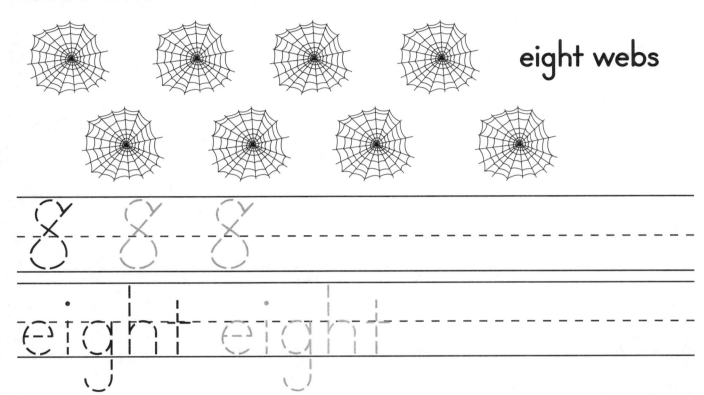

eight webs

Circle and color eight spiders.

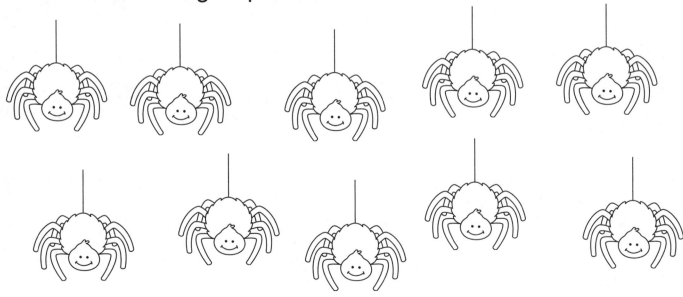

Writing and Counting to Nine

Count and color the hats. Practice writing the numeral and number word.

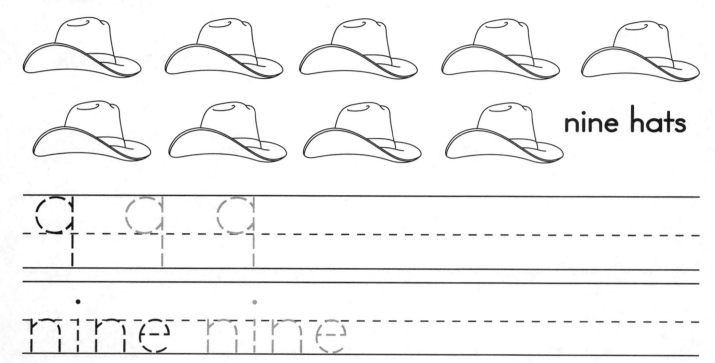

nine hats

Circle and color nine horses.

Name _____

Writing and Counting to Ten

Count and color the bats. Practice writing the numeral and number word.

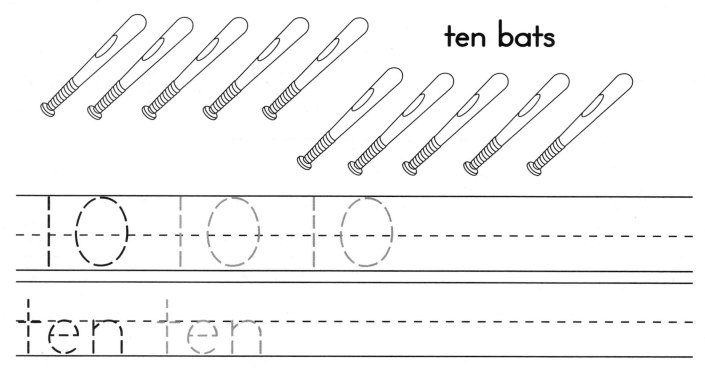

ten bats

10 10 10

ten ten

Circle and color ten balls.

Counting to Five

Count the number of objects in each row.
Circle the correct numeral.

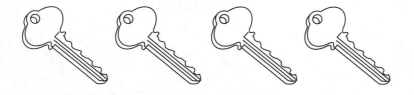 1 2 3 4 5

   1 2 3 4 5

 1 2 3 4 5

  1 2 3 4 5

  1 2 3 4 5

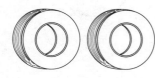   1 2 3 4 5

Counting to Ten

Count the number of objects in each row.
Circle the correct numeral.

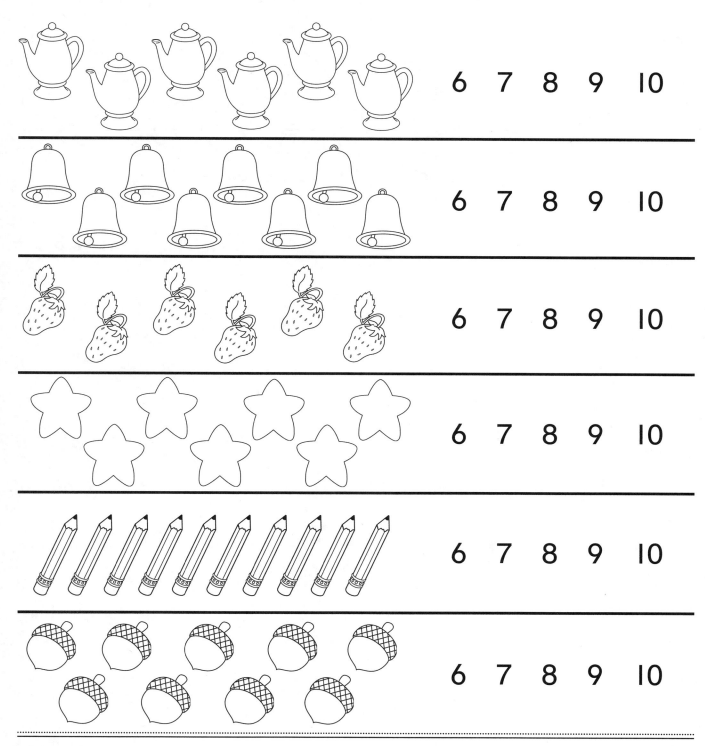

6 7 8 9 10

6 7 8 9 10

6 7 8 9 10

6 7 8 9 10

6 7 8 9 10

6 7 8 9 10

Name _____

Counting 1–15

Connect the dots from **1–15**. Color the completed picture.
Start at the star.

14●

15●

13●

●2

●3

12●

●4

11●

●5

●
10

●
9

●
8

●
7

●
6

CD-104638 • © Carson-Dellosa

Name _____

Counting 1–20

Connect the dots from **1–20**. Color the completed picture.
Start at the star.

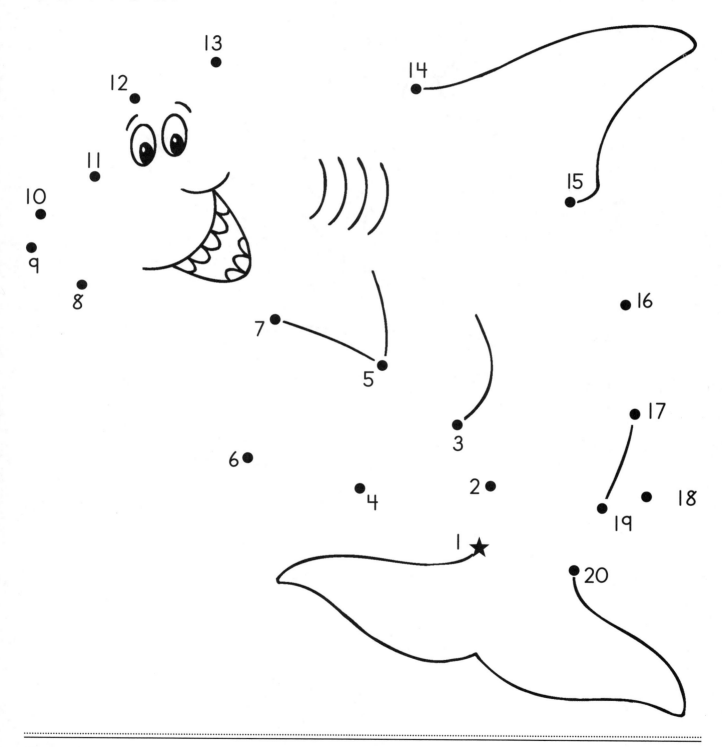

Counting 1–25

Connect the dots from **1–25**. Color the completed picture.
Start at the star.

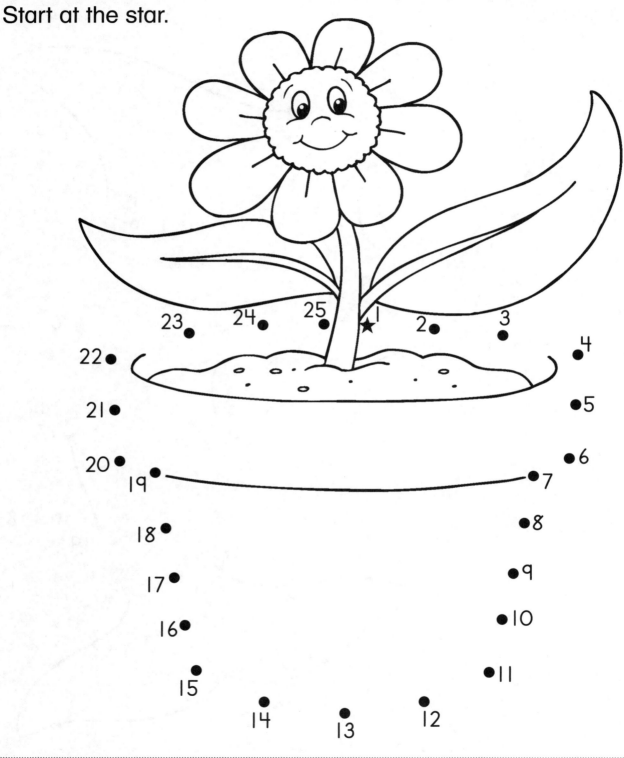

Counting 1–30

Connect the dots from **1–30**. Color the completed picture.
Start at the star.

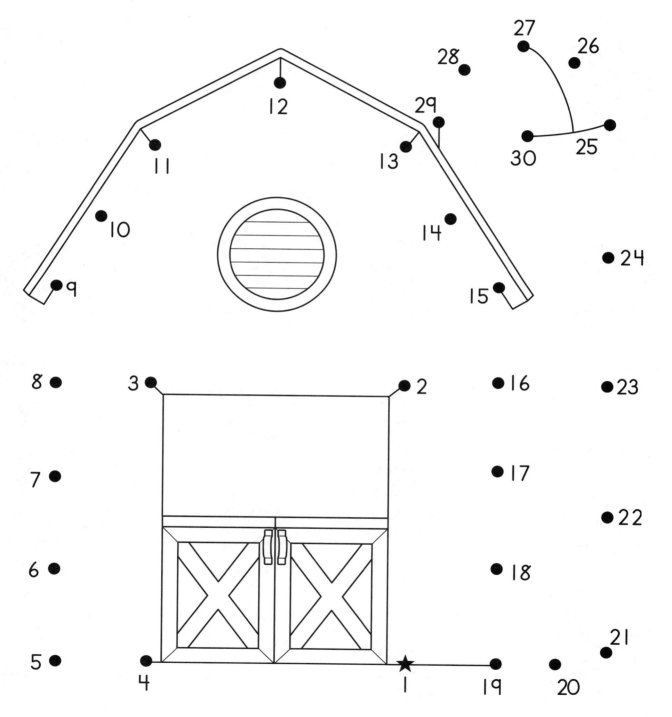

Name _____

Recognizing First

Circle what happened **first** in each row.

  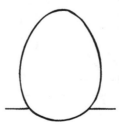 

Name _____

Recognizing Last

Circle what happened **last** in each row.

Name _____

Sequencing Events

Cut out the pictures below and paste them in order from first to last.

1	2
3	4

CD-104638 • © Carson-Dellosa

Name _____

Sequencing Events

Cut out the pictures below and paste them in order from first to last.

1	2
3	4

Sequencing Events

Cut out the pictures below and paste them in order from first to last.

1	2
3	4

Name _____

Sequencing Events

Cut out the pictures below and paste them in order from first to last.

1	2
3	4

Sequencing Events

Cut out the pictures below and paste them in order from first to last.

1	2
3	4

Name _____

Sequencing Events

Cut out the pictures below and paste them in order from first to last.

1	2
3	4

Tracing Squares

Trace the dotted lines to complete the **squares**. Color each square a different color.

Tracing Circles

Trace the dotted lines to complete the **circles**. Color three circles black. Color three circles red.

Tracing Rectangles

Trace the dotted lines to complete the **rectangles**. Color each rectangle a different color.

Tracing Triangles

Trace the dotted lines to complete the **triangles**. Make each triangle into a slice of pizza. Color each slice of pizza.

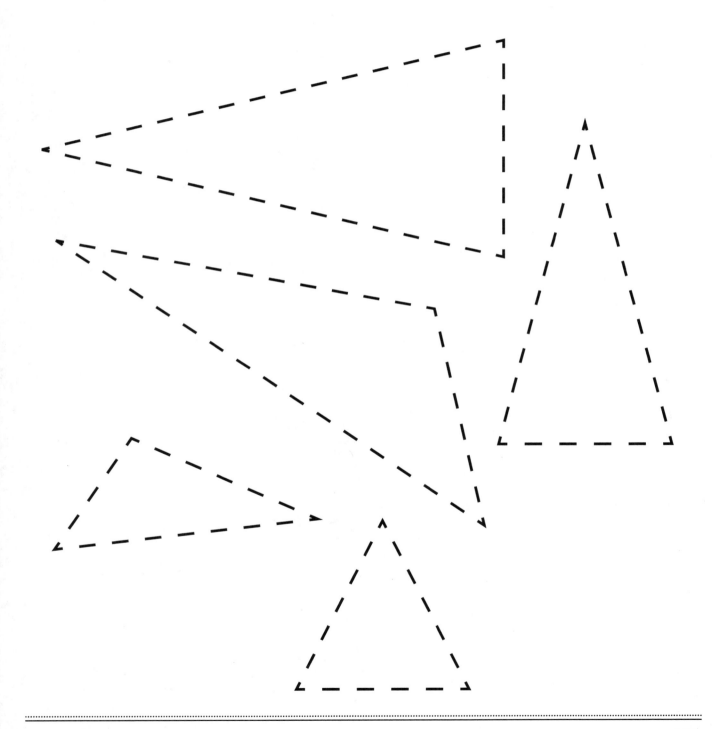

Tracing Ovals

Trace the dotted lines to complete the **ovals**. Color the ovals red.

Tracing Hexagons

Trace the dotted lines to complete the **hexagons**. Color the hexagons yellow.

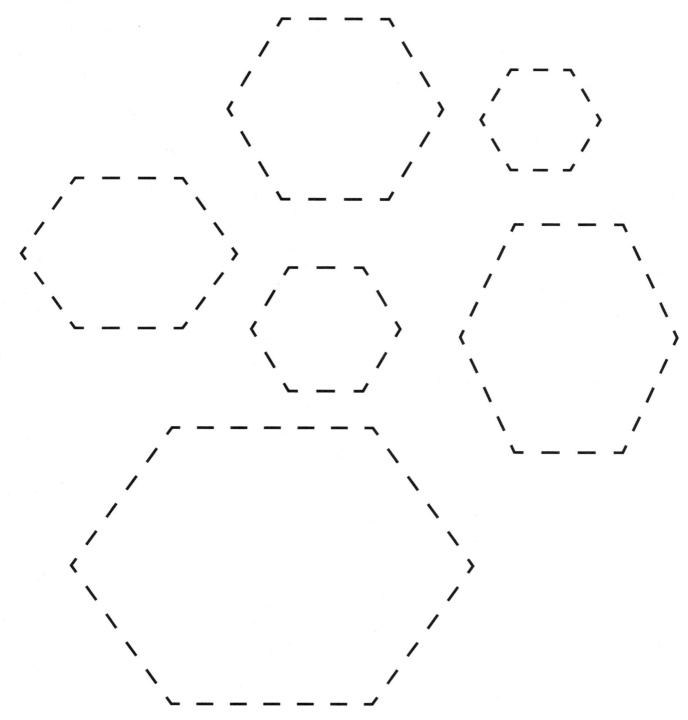

Tracing Rhombuses

Trace the dotted lines to complete the **rhombuses**. Color the rhombuses yellow.

Name _____

Recognizing Squares

Color all of the **squares** yellow. Draw a blue X over all of the other shapes.

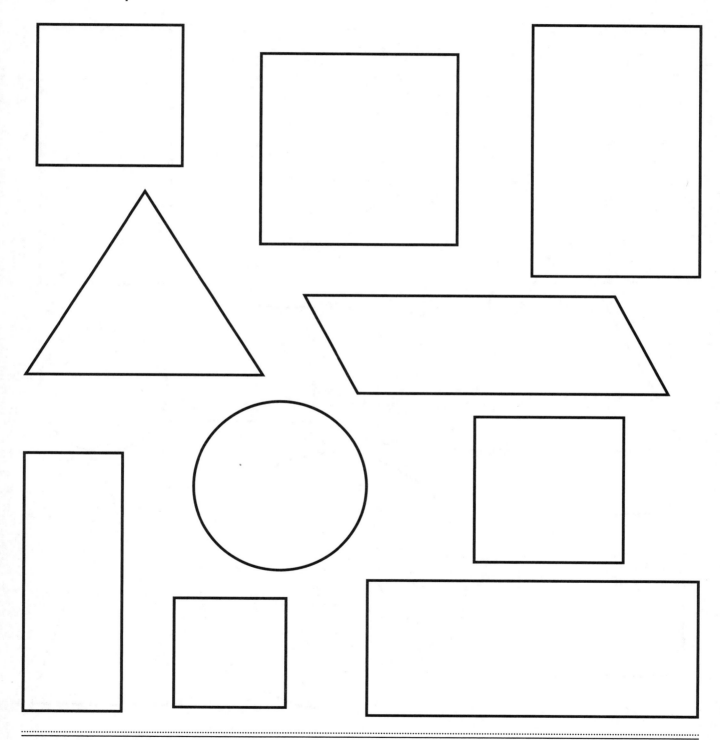

Recognizing Circles

Color all of the **circles** green. Draw a red X over all of the other shapes.

Recognizing Rectangles

Color all of the **rectangles** orange. Draw a brown X over all of the other shapes.

Recognizing Triangles

Color all of the **triangles** brown. Draw a pink X over all of the other shapes.

Name _____

Recognizing Ovals

Color all of the **ovals** red. Draw a black X over all of the other shapes.

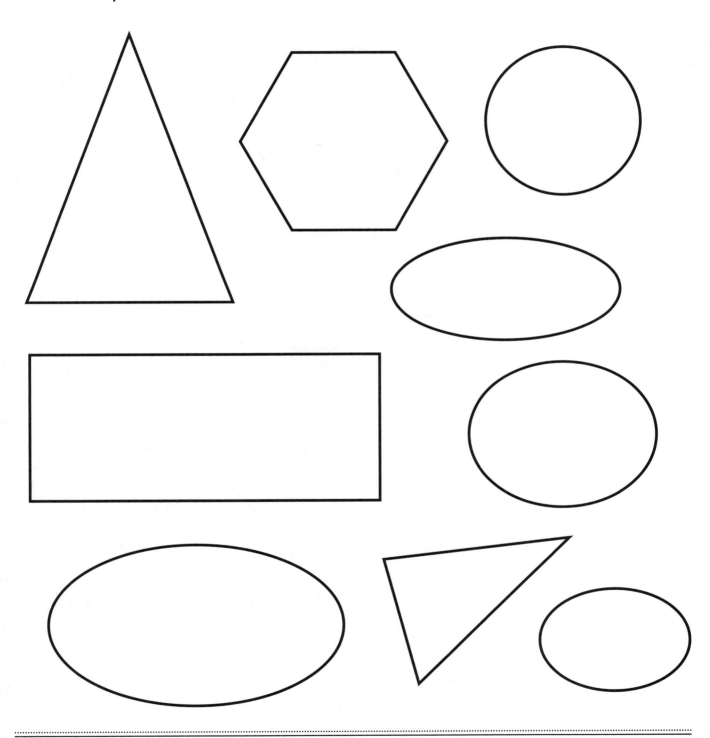

Name _____

Recognizing Hexagons

Color all of the **hexagons** red. Draw a green X over all of the other shapes.

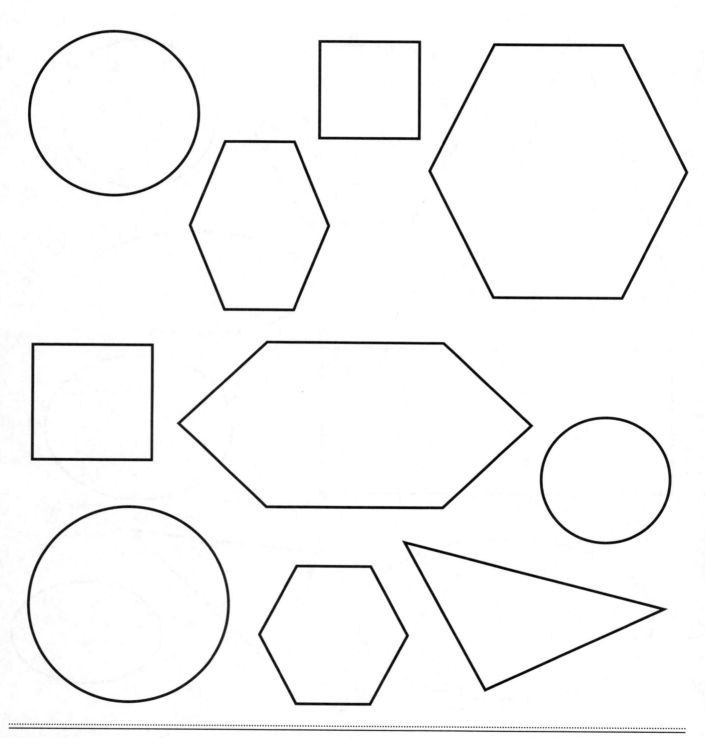

CD-104638 • © Carson-Dellosa

Name _____

Recognizing Rhombuses

Color all of the **rhombuses** yellow. Draw a purple X over all of the other shapes.

Name _____

Drawing Shapes

Practice drawing each shape.

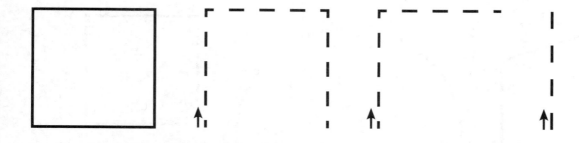

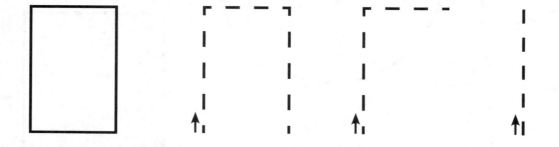

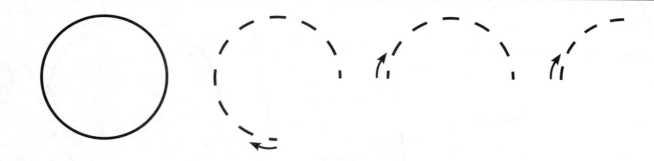

Name _____

Drawing Shapes

Practice drawing each shape.

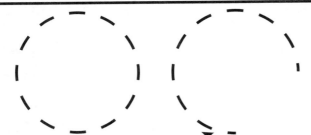

Name _____

Drawing Shapes

Practice drawing each shape.

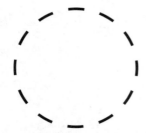

_____

_____

_____

_____

Name _____

Matching Shapes

Draw a line to match the shapes that are the same. Color the matching shapes the same color.

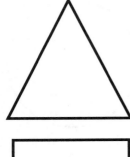

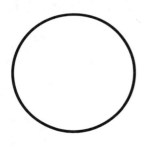
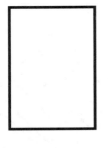
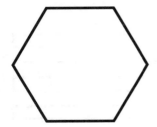

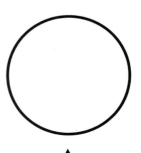
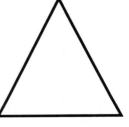
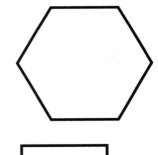
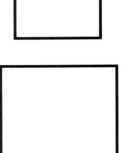

Matching Shapes

Draw a line to match the shapes that are the same. Color the matching shapes the same color.

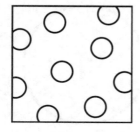
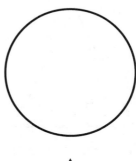
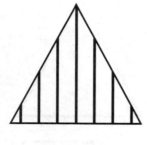
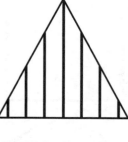
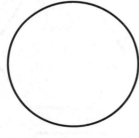
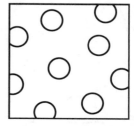
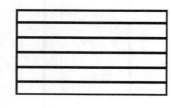

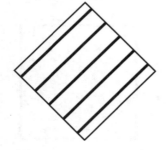
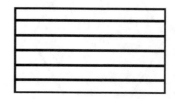

Name _____

Draw and Print Myself

· ·

Draw a picture of yourself.

My name is

_____

- -

_____.

Draw and Print My Telephone Number

Draw a picture of yourself talking on the telephone.

I know my telephone number. It is

_____

- -

_____ .

Name _____

Draw and Print My Home

Draw a picture of your home.

My address is

_____

_ _

_____ .

Name _____

Draw and Print a Spring Activity

Draw a picture of something you like to do in spring.

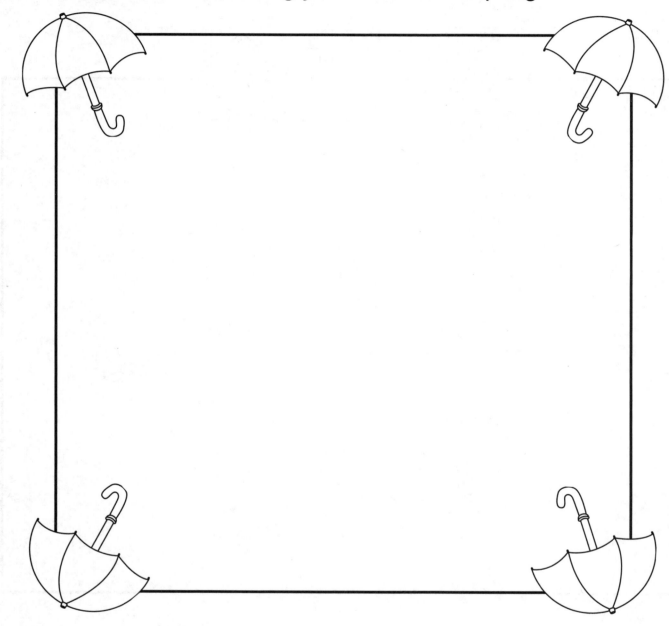

My favorite spring month is

_____

_ _

_____ .

Name _____

Draw and Print a Summer Activity

Draw a picture of something you like to do in summer.

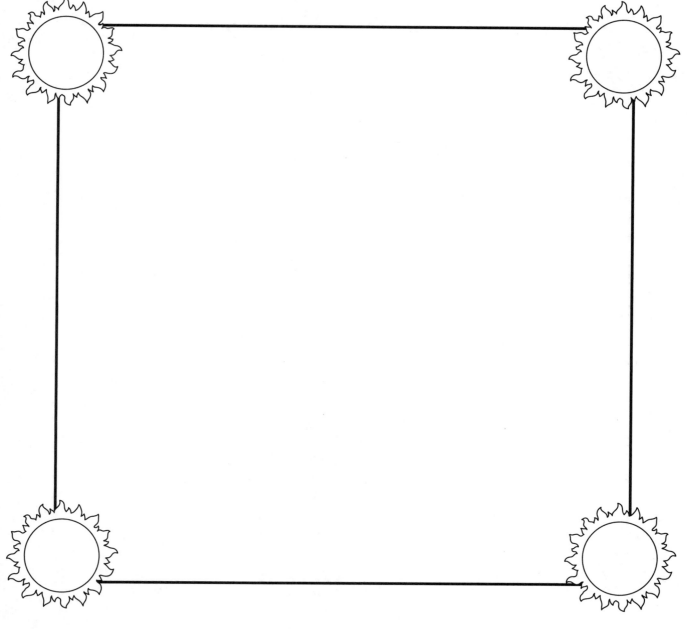

My favorite summer month is

_____

_ _

_____.

Name _____

Draw and Print an Autumn Activity

Draw a picture of something you like to do in autumn.

My favorite autumn month is

_____

- -

_____ .

Name _____

Draw and Print a Winter Activity

Draw a picture of something you like to do in winter.

My favorite winter month is

_____

_ _

_____.

Draw and Print My Favorite Food

Draw a picture of your favorite food.

My favorite food is

_____

— —

_____ .

Name _____

Draw and Print My Favorite Place

Draw a picture of yourself at your favorite place.

My favorite place is
_____

— —

_____ .

Name _____

Draw and Print My Favorite Animal

Draw a picture of your favorite animal.

My favorite animal is
_____

– –

_____ .

Name _____

Draw and Print My Favorite Book

Draw a character from your favorite book.

My favorite book is
_____

- -

_____ .

Name _____

Draw and Print My Family

Draw a picture of something your family likes to do.

My family likes to

_____

- -

_____ .

Congratulations!

receives this award for

Signed _____

Date _____

three	two	one	zero
seven	six	five	four
eleven	ten	nine	eight
fifteen	fourteen	thirteen	twelve

© CD

sixteen   seventeen   eighteen   nineteen

twenty   twenty-one   twenty-two   twenty-three

twenty-four   twenty-five   twenty-six   twenty-seven

twenty-eight   twenty-nine   thirty   thirty-one

thirty-two	thirty-three	thirty-four	thirty-five
© CD	© CD	© CD	© CD
thirty-six	thirty-seven	thirty-eight	thirty-nine
© CD	© CD	© CD	© CD
forty	forty-one	forty-two	forty-three
© CD	© CD	© CD	© CD
forty-four	forty-five	forty-six	forty-seven
© CD	© CD	© CD	© CD

forty-eight

forty-nine

fifty

sixty

seventy

eighty

ninety

one hundred

>

+

d

−

b

=

c

<

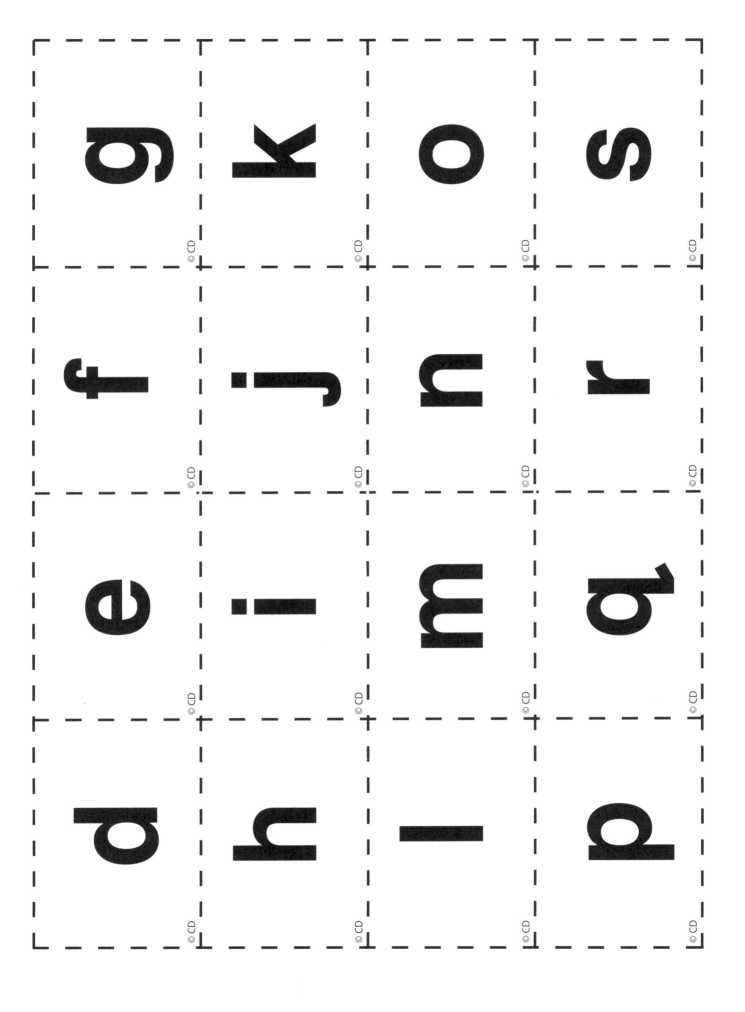

F   B   X   T

G   C   Y   U

H   D   Z   V

I   E   A   W

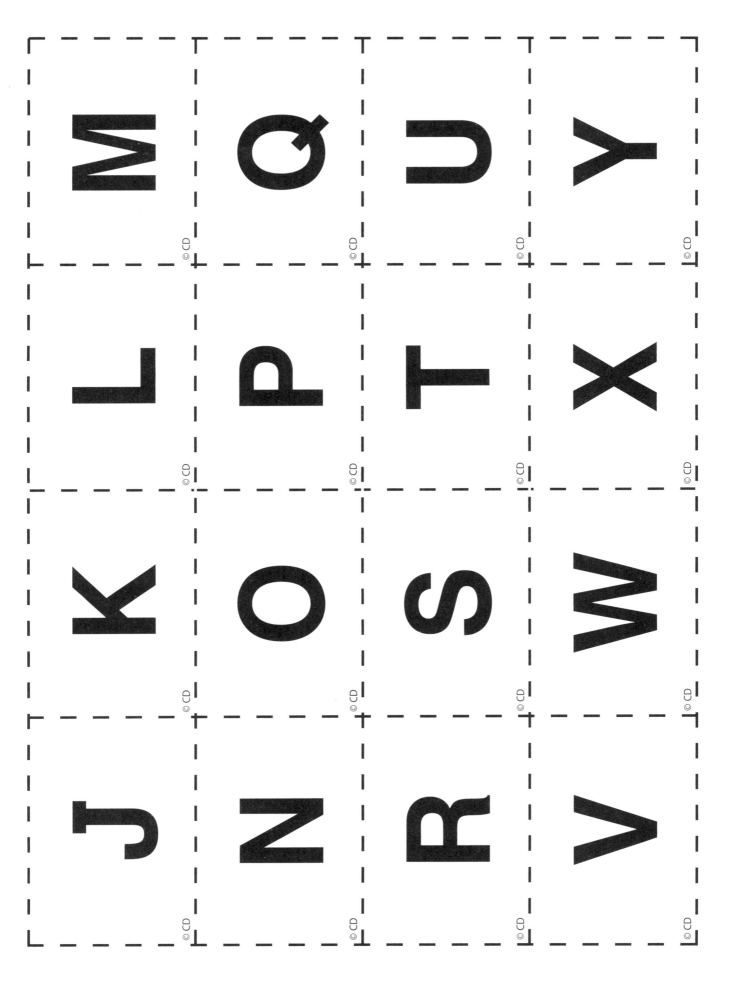

2	1	2	3
4	5	6	7
8	9	10	11
12	13	14	15

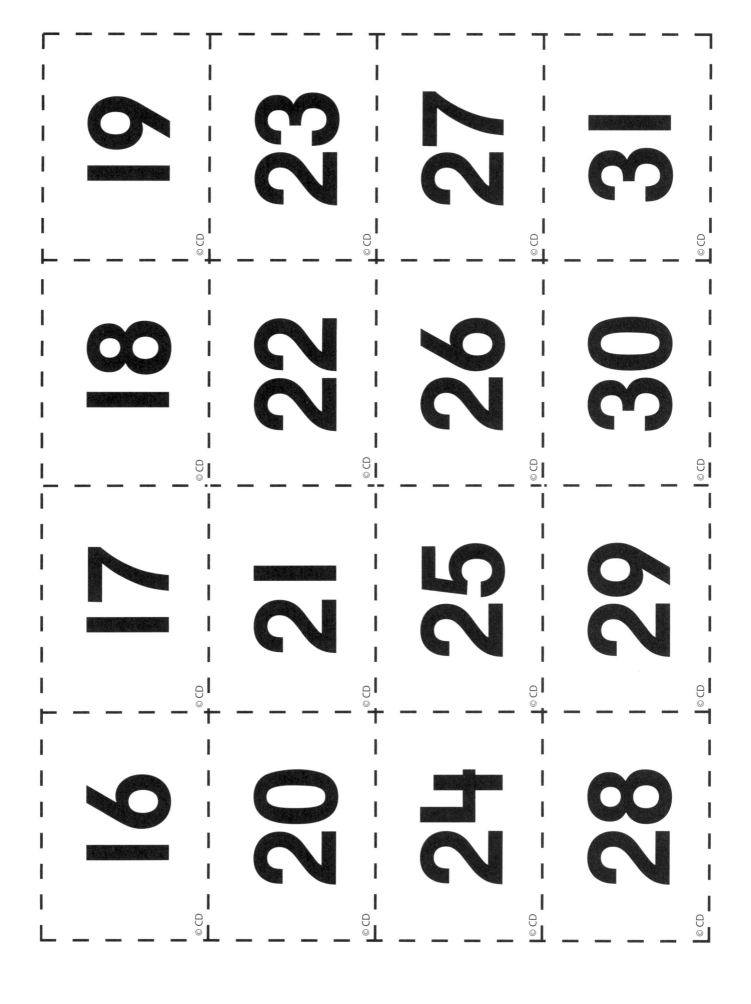

32 33 34 35

36 37 38 39

40 41 42 43

44 45 46 47

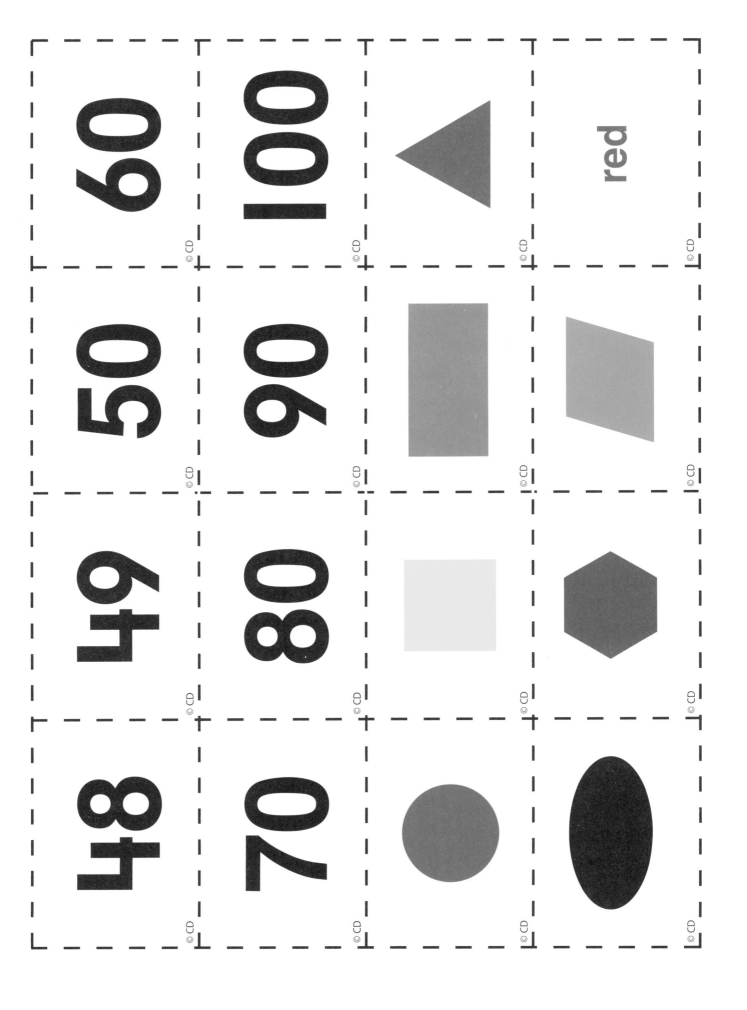

yellow

brown

triangle

green

orange

rectangle

rhombus

blue

square

hexagon

black

circle

oval